Geotechnical Engineering

Projects

Two Comprehensive Design Projects

Arzhang Zamani, Ph.D.

STRUCTURE GATE

Geotechnical Engineering Projects

Copyright © 2018 by STRUCTURE GATE

ISBN (978-1-948135-04-7)

Printed in the USA

Disclaimer

Dedication

To Farzaneh and Mehrdad

With Love and Passion

Table of Contents

Introduction

The purpose of this book is to familiarize the reader with process of designing a foundation for bridge pier and design of retaining wall.

This book is useful for civil engineering students for their geotechnical and foundation engineering projects. It can be a good companion with geotechnical engineering textbooks.

Each section contains defining the problem statement and aim of the project. The introduction is followed by the step by step procedure to design and meet the requirements for the project.

First chapter includes the design of a foundation for bridge pier, calculations for settlement with different methods, and structural design of reinforcement for the foundation.

Second chapter covers retaining wall design with a real-life criterion by using geogrid and segmental blocks. Detailing for final design is presented in the appendix.

I hope you enjoy reading this book.

Live with Passion!

Arzhang Zamani

Chapter 1

Pier Foundation Design

1.1 "Problem Statement"

You have been tasked to assist the structural engineer for a bridge replacement project. Your task is to design the size of the footing for a bridge pier, based on the allowable stress design method. A soil test boring was drilled at the location of the bridge pier. The soil test boring log has shown that the soil is fairly uniform at the pier location and consists of a layer of clay overlying a five-layer system of silty sand, well graded sand, uniform sand, sandy silt, and well-graded sand. Detailed information about each layer is given in the attached soil test boring log and in the table below. Groundwater table was encountered at the depth of 25 feet after a stabilization period during the subsurface exploration.

Neglect correction factors for borehole diameter, sampler correction, and correction for rod length

Design an appropriate footing for bridge pier, based on ASD method.

"Fundamental Characteristics"

- Scour is not an issue at the site

- Frost penetration depth is 3.0 ft.

- Axial Load = 3200 kips

- Shear Load = 100 kips

- Moment in Z-direction = 800 kips-ft

- Moment in y-direction = 1000 kips-ft

- Use Schmertmann's (1975) and Kulhawy and Mayne (1990) correlation between SPT and φ' and round to the nearest degree

- The increase in stress in each layer should be estimated using 2:1 method

- Settlement analysis should be based on the Hough Method, the Schmertmann's method, and the Burland & Burbidge's method.

- Bridge pier is 3 ft. by 3ft.

- Ground Water table was encountered at the depth of 25 ft. after a stabilization period during the subsurface exploration. $\gamma_w = 62.4 \ pcf$

Table 1 Relation between q_c and N_{60} in tsf

Soil Type	q_c/N_{60}
Silts, sandy silts, slightly cohesive silt-sand	2
Clean, fine to medium sands and silty sands and gravels	3.5
Fine to coarse sands and sands with gravel	4
Sandy gravel and gravel	6

Table 2 Soil properties in each layer

Depth (ft)	Soil Description	Wet unit weight (pcf)	Saturated Unit Weight (pcf)
0.25-5.5	Lean Clay	120	--
5.5-16	Silty SAND & GRAVEL	125	--
16-40	Well graded fine to coarse SAND	120	130
40-52	Uniform medium SAND	115	125
52-76	SANDY SILT	110	120
76-100	Well graded fine to coarse SAND	120	130

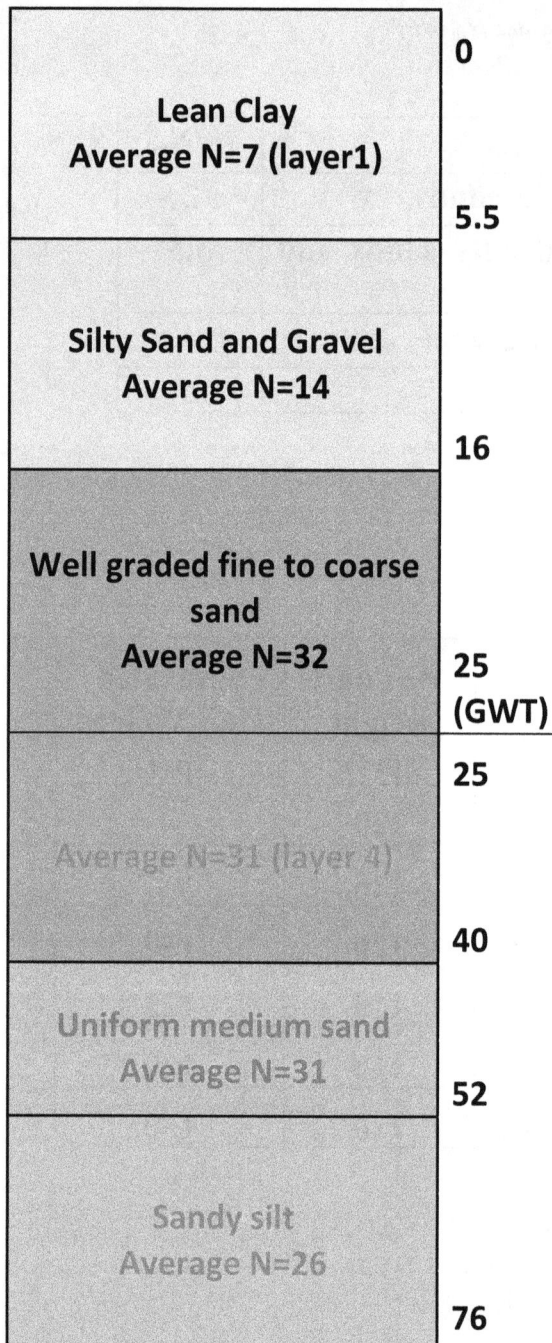

Figure 1 Layers and average N value.

Lean Clay Average N=7 (layer1)	0
	5.5
Silty Sand and Gravel Average N=14	
	16
Well graded fine to coarse sand Average N=32	
	25 (GWT)
Average N=31 (layer 4)	25
	40
Uniform medium sand Average N=31	
	52
Sandy silt Average N=26	
	76

Problem Statement:

Using the information provided above, design the size of the footing for the interior bridge pier, based on the allowable stress design method. You should consider both settlement and bearing capacity, and check for overturning and sliding. You should provide the structural engineer with a formal typed report that will include all your analyses and final size of the footing, including the thickness of the footing, reinforcement, and CAD drawing of the footing. You should also include all your design assumptions in your report and explain why you chose that footing size.

Plan view of pier:

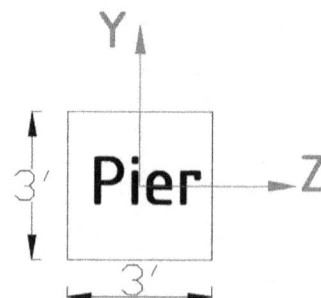

1.2 Abstract

This project is about designing a shallow foundation for bridge pier. The most reliable equations and procedures have been used to evaluate and determine the parameters of design. Design is in accordance with AASHTO LRFD 2012 and the text book of The Geotechnical Engineering course. The footing is designed to satisfy the requirement for all applicable factors. The settlement analysis according the tests is performed by using three methods. Bearing capacity is calculated and compared with regarding factor of safety of 3. Required foundation thickness has been designed to meet the requirement for one-way shear and punching shear. Required reinforcement has been calculated and placed in the footing according to ACI 318-14. All the calculations related to this foundation except ACI sections are based on ASD method.

1.3 Introduction:

In this project by using the materials that we learned from the Geotechnical Engineering course, a footing is designed for a bridge pier.

1- The outline for this design is as following:

2- Extracting all possible information from boring log and result of SPT test.

3- Finding the mechanical properties of soil in each layer according to test result and recommended equations

4- Performing Analysis for first estimate of footing size according to given loads.

5- Checking the settlement using one of the three proposed methods for settlement analysis

6- Increasing the footing width to be in the acceptable limit of settlement

7- Checking all of three methods for settlement to feel certain that all of them are within the maximum allowable settlement

8- Performing the bearing capacity calculation to double check that, the foundation is safe against applied loads.

9- Finding the required thickness of foundation to satisfy one way and punching shear

10- Reinforcement design

11- Detailing and final drawings

It is necessary to mention that all the requirements must be met to finalize the

design of foundation for this bridge pier.

1.4 Phase I: Subsurface Exploration Analysis

It is necessary to describe the mechanical characteristics of soil by using provided subsurface exploration tests. (PCA 1993)

$$\eta_h = 79\% \; Hammer \; Efficiency$$

$\eta_B = 1 \; \eta_s = 1 \; \eta_R = 1$

Layer 1: (0.25-5.5ft)

(N)ave = 7

$$N_{60} = \frac{N_{ave}(\eta_h)(\eta_B)(\eta_s)(\eta_R)}{60} = \frac{7(79)(1)(1)(1)}{60} = 9$$

Arzhang Zamani

(Kulhawy and Mayne,1990)

$$\sigma'_0 = 120\left(\frac{5.5}{2}\right) = 330\ pcf\ \ P_a = 2{,}000\ pcf$$

$$\varphi' = \tan^{-1}\left(\frac{N_{60}}{12.2 + 20.3\left(\frac{\sigma'_0}{P_a}\right)}\right)^{0.34}$$

$$\varphi' = \tan^{-1}\left(\frac{9}{12.2 + 20.3\left(\frac{330}{2000}\right)}\right)^{0.34} = 40\ °$$

$$C_N = \left(\frac{1}{\frac{\sigma'_0}{2000}}\right)^{0.5} = 2.46$$

$$(N_1)_{60} = C_N N_{60} = 22$$

Layer 2: (5.5-16 ft)

(N)ave = 14

$$N_{60} = \frac{N_{ave}(\eta_h)(\eta_B)(\eta_s)(\eta_R)}{60} = \frac{14(79)(1)(1)(1)}{60} = 18$$

(Kulhawy and Mayne, 1990)

$$\sigma'_0 = 120(5.5) + 125(5.25) = 1316\ pcf\ \ P_a = 2{,}000\ pcf$$

$$\varphi' = \tan^{-1}\left(\frac{18}{12.2 + 20.3\left(\frac{1316}{2000}\right)}\right)^{0.34} = 42°$$

$$C_N = \left(\frac{1}{\frac{\sigma'_0}{2000}}\right)^{0.5} = 1.23$$

$$(N_1)_{60} = C_N N_{60} = 22$$

Layer 3: (16-25 ft)

(N)ave = 32

$$N_{60} = \frac{N_{ave}(\eta_h)(\eta_B)(\eta_s)(\eta_R)}{60} = \frac{32(79)(1)(1)(1)}{60} = 42$$

(Kulhawy and Mayne,1990)

$$\sigma'_0 = 120(5.5) + 125(10.5) + 120(4.5) = 2513 \quad P_a = 2,000 \, pcf$$

$$\varphi' = \tan^{-1}\left(\frac{N_{60}}{12.2 + 20.3\left(\frac{\sigma'_0}{P_a}\right)}\right)^{0.34}$$

$$\varphi' = \tan^{-1}\left(\frac{42}{12.2 + 20.3\left(\frac{2513}{2000}\right)}\right)^{0.34} = 46°$$

$$C_N = \left(\frac{1}{\frac{\sigma'_0}{2000}}\right)^{0.5} = 0.89$$

$$(N_1)_{60} = C_N N_{60} = 37$$

Layer 4: (25-40 ft)

(N)ave = 31

$$N_{60} = \frac{N_{ave}(\eta_h)(\eta_B)(\eta_s)(\eta_R)}{60} = \frac{31(79)(1)(1)(1)}{60} = 41$$

(Kulhawy and Mayne,1990)

$$\sigma'_0 = 120(5.5) + 125(10.5) + 120(9) + (130 - 62.4)(7.5) = 3561.5$$

$$P_a = 2,000 \ pcf$$

$$\varphi' = \tan^{-1}\left(\frac{41}{12.2 + 20.3\left(\frac{3561.5}{2000}\right)}\right)^{0.34} = 43\ °$$

$$C_N = \left(\frac{1}{\frac{\sigma'_0}{2000}}\right)^{0.5} = 0.75$$

$$(N_1)_{60} = C_N N_{60} = 31$$

Layer 5: (40-52 ft)

(N)ave = 31

$$N_{60} = \frac{N_{ave}(\eta_h)(\eta_B)(\eta_s)(\eta_R)}{60} = \frac{31(79)(1)(1)(1)}{60} = 41$$

(Kulhawy and Mayne,1990)

$$\sigma'_0 = 120(5.5) + 125(10.5) + 120(9) + (130 - 62.4)(15) +$$

$$(125 - 62.4)6 = 4442 \quad P_a = 2,000 \, pcf$$

$$\varphi' = \tan^{-1}\left(\frac{41}{12.2 + 20.3\left(\frac{4442}{2000}\right)}\right)^{0.34} = 42°$$

$$C_N = \left(\frac{1}{\frac{\sigma'_0}{2000}}\right)^{0.5} = 0.67$$

$$(N_1)_{60} = C_N N_{60} = 28$$

Layer 6: (52-76 ft)

(N)ave = 26

$$N_{60} = \frac{N_{ave}(\eta_h)(\eta_B)(\eta_s)(\eta_R)}{60} = \frac{26(79)(1)(1)(1)}{60} = 34$$

Arzhang Zamani

(Kulhawy and Mayne,1990)

$$\sigma'_0 = 120(5.5) + 125(10.5) + 120(9) + (130 - 62.4)(15) +$$

$$(125 - 62.4)12 + (120 - 62.4)12 = 5508.9 \quad P_a = 2,000\,pcf$$

$$\varphi' = \tan^{-1}\left(\frac{34}{12.2 + 20.3\left(\frac{5508.9}{2000}\right)}\right)^{0.34} = 38°$$

$$C_N = \left(\frac{1}{\frac{\sigma'_0}{2000}}\right)^{0.5} = 0.6$$

$$(N_1)_{60} = C_N N_{60} = 20$$

Layer 7: (76-100 ft)

(N)ave = 46

$$N_{60} = \frac{N_{ave}(\eta_h)(\eta_B)(\eta_s)(\eta_R)}{60} = \frac{46(79)(1)(1)(1)}{60} = 61$$

(Kulhawy and Mayne,1990)

$$\sigma'_0 = 120(5.5) + 125(10.5) + 120(9) + (130 - 62.4)(15) +$$

$$(125 - 62.4)12 + (120 - 62.4)24 + (130 - 62.4)12.5 = 7045.1$$

$$P_a = 2,000\,pcf$$

20

$$\varphi' = \tan^{-1}\left(\frac{61}{12.2+20.3\left(\frac{7045.1}{2000}\right)}\right)^{0.34} = 42°$$

$$C_N = \left(\frac{1}{\frac{\sigma'_0}{2000}}\right)^{0.5} = 0.53$$

$$(N_1)_{60} = C_N N_{60} = 32$$

As can be seen, for all the layers the main characteristics of soli have been driven. In the next steps by utilizing these characteristics behavior of layers against various conditions are evaluated.

1.5 Phase II: Preliminary Design

To start process of design, we must have an approximate estimate for the size of footing.

In the beginning, it is assumed that the bearing capacity must be met, then the settlement will be checked against that size. Most probably settlement will be governing for this problem. Therefore, after the initial estimate for footing size according to soil bearing capacity. The size of footing will be added incrementally to satisfy settlement criteria.

Arzhang Zamani

1.6 Important Note

The depth of foundation should be greater than 3 ft. which is the frost depth. According to the boring log, it is more efficient to choose the depth of foundation at 5.5 ft to bear the foundation on the sand layer. This action helps to improve general performance of footing. This depth is reachable by contractor and the problems that arose while having clay on the first layer will be diminished.

Assume:

$D_f = 5.5\ ft.$ Square footing

Layer 2:

$\varphi' = 42\ °$

FS = 3

$q_{all} = \dfrac{Q_{all}}{B^2}$ $q_{all} = \dfrac{q_u}{FS}$

$N_q = 85.38$ $N_\gamma = 155.55$ $\gamma = 0.125\ kcf$

$q_{all} = \dfrac{1}{3}\left(qN_qF_{qs}F_{qd} + \dfrac{1}{2}\gamma BN_\gamma F_{\gamma s}F_{\gamma d}\right)$

$$q = 0.120(5.5) = 0.66\ kip/ft^2$$

$Q_{all} = 3,200\ kips * 1.3$ (only approximate estimate effect of moments)

$F_{qs} = 1.9$ $F_{\gamma s} = 0.6$ $F_{qd} = 1 + 2\tan(42)\,(1 - \sin(42))^2 * \dfrac{D}{B}$ $F_{\gamma d} = 1$

Solve for B, B=8.55 ft. USE B=9ft.

Let's check the settlement.

Since, we have eccentric load. It is important to meet requirement for settlement.

According to created spreadsheet the settlement analysis is checked for Hough method. Then, after meeting the requirement. The detailed analyses for all of three methods and bearing capacity are going to be used to finalize the dimension.

1.7 Settlement analysis based on Hough method:

All the criterion for using Hough's method are applied to calculate the settlement based on the problem conditions.

q=	22222.22	lbs.					
B=	12	ft					
Characteristic	Layer2	Layer3	Layer4	Layer5	Layer6	Layer7	
H(ft)	10.5	9	15	12	24	24	
σ'0(middle)(psf)	1316	2513	3562	4442	5509	7045	
Z(ft)	5.25	15	27	40.5	58.5	87.5	
Δσ(v)	10754.04	4390	2104	1161	643.8	323.2	
C'	85	90	130	115	60	95	
N1(60)	22	37	31	28	20	32	
							sum:
ΔH(ft)	0.12	0.04	0.023	0.01	0.019	0.004	0.22
ΔH(in)	1.43	0.53	0.28	0.13	0.23	0.06	2.65
							NG

By increasing the dimension size and considering the 1.5 in. limit for the settlement.

The 28 ft width is acceptable. Since, we have also eccentric moment, and 30 is a factor of 5. The 30 ft. is selected.

q=	3555.56	lbs.					
B=	30	ft					
Characteristic	Layer2	Layer3	Layer4	Layer5	Layer6	Layer7	
H(ft)	10.5	9	15	12	24	24	
σ'0(middle)(psf)	1316	2513	3562	4442	5509	7045	
Z(ft)	5.25	15	27	40.5	58.5	87.5	
Δσ(v)	2575.32	1580	984.9	643.8	408.6	231.8	
C'	85	90	130	115	60	95	
N1(60)	22	37	31	28	20	32	
							sum:
ΔH(ft)	0.059	0.02	0.012	0.006	0.012	0.003	0.12
ΔH(in)	0.70	0.25	0.15	0.07	0.15	0.04	1.37
							Good

It can be seen from this example that settlement is governing. Therefore, by checking the real settlement according to suggested three methods and comparing with the allowable deflection, the width will be found.

1.8 Analyzing Effect of eccentricity:

$$e_{B(z)} = \frac{M_y}{Q_{ult}} = \left(\frac{1000}{3200}\right) = 0.3125 \, ft.$$

$$e_{L(y)} = \frac{M_z}{Q_{ult}} = \left(\frac{800}{3200}\right) = 0.25 \, ft.$$

Therefore,

$$\frac{e_{B(z)}}{B} = 0.01 \qquad \frac{e_{L(y)}}{B=L} = 0.0083$$

Both of previous ratios are less than $\left(\frac{1}{6}\right)$. Therefore, case 4 is valid.

$$A' = L_2 B + \frac{1}{2}(B + B_2)(L - L_2)$$

For finding B_2 & L_2. According to Fig 3.23:

$$\frac{L_2}{L} = 0.9 \qquad \frac{B_2}{B} = 0.9 \quad B = 30 ft$$

$$A' = L_2 B + \frac{1}{2}(B + B_2)(L - L_2) = 0.995 B^2 = 895.5 \, ft^2$$

$$\frac{A'}{B} = \frac{895.5}{30} = 29.85 \, ft$$

So, based on calculation and considering effect of eccentricity.

The effective length is 30 x 29.85 ft.

Therefore, it is possible to neglect effect of eccentricity for this problem according to the applied load, moment and size of footing.

All the calculations will be performed by considering a square footing with 30 ft length.

1.9   Settlement analysis based on Schmertmann's method:

$$S_e = C_1 C_2 (\bar{q} - q) \sum (\frac{I_z}{E_s}) \Delta z$$

$$C_1 = 1 - 0.5(\frac{q}{\bar{q} - q})$$

$$\bar{q} = \frac{3200}{30^2} = 3.556 \frac{kip}{ft^2}$$

$$q = 5.5(120) * \left(\frac{1}{1000}\right) = 0.66 \frac{kip}{ft^2}$$

$$Z_1 = 0.5B = 15ft$$

$$C_1 = 1 - 0.5 \left(\frac{q}{\bar{q} - q}\right) = 0.886$$

$$q'_{z1} = 120(5.5) + 125(10.5) + 120(15 - 10.5) = 2.5125 \frac{kip}{ft^2}$$

$$I_{z(m)} = 0.5 + 0.1 \sqrt{\frac{\overline{q} - q}{q'_{z1}}} = 0.607$$

At Z=0 $I_z = 0.1$

$$Z_2 = 2B = 60 \, ft$$

Soil Profile:

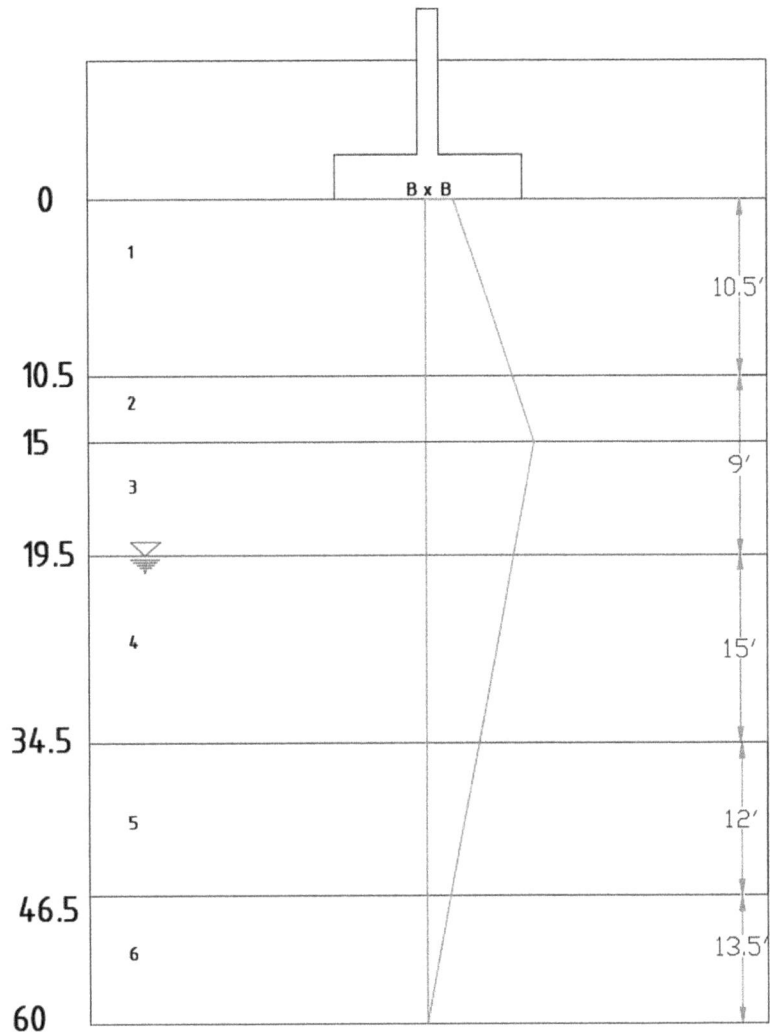

Figure 2 Soil profile and Influence diagram

I_z **variation to depth of 2B=60 ft.**

The required values are obtained and displayed on the next page.

C1	0.886
C2	1.4
q bar	3.59
q	0.66

Necessary values for Schmertmann's equation:

Schmertmann's method:

	E(ksf)	delta h(ft)	Iz	Iz(deltaH)/E
1	315	10.5	0.3535	0.012
2	840	4.5	0.5793	0.003
3	840	4.5	0.5242	0.003
4	820	15	0.4047	0.007
5	717.5	12	0.2391	0.004
6	340	13.5	0.0828	0.003
			sigma	0.032
			ft	in
	DELTA		0.12	1.44
			Good	

Settlement is 1.42 in based on Schmertmann's method. It is very close and less than allowable limit.

1.10 Settlement analysis based on Burland and Burbidge's method:

This method is for calculating elastic settlement of sandy soil using field SPT, N_{60} values.

$$B_R = 1$$

$$Z' = 1.4(B)^{0.75} = 1.4(30)^{0.75} = 17.946 \text{ ft}$$

$$\alpha_1 = 0.14 \quad \alpha_2 = \frac{1.71}{N_{60}^{1.4}} \quad \alpha_3 = \frac{H}{Z'}\left(2 - \frac{H}{Z'}\right) = 1$$

$$S_e = \alpha_1 \alpha_2 \alpha_3 \left(\frac{1.25}{0.25 + 1}\right)^2 (B)^{0.7}\left(\frac{q'}{Pa}\right)$$

$$q' = \left(\frac{320}{30^2}\right) - 0.12(5.5) = 2.896\frac{kip}{ft^2}$$

$$Pa = 2\frac{kip}{ft^2}$$

N_{60} is the average of N_{60} in the depth of Z'.

$$N_{60} = \left(\frac{10.5}{17.946}\right) * 18 + \left(\frac{7.448}{17.946}\right) 42 = 28$$

$$\alpha_2 = \frac{1.71}{N_{60}^{1.4}} = \frac{1.71}{28^{1.4}} = 0.01611$$

$$S_e = 0.14 * (0.01611)(30)^{0.7} \left(\frac{2.896}{2}\right) = 0.035 \, ft = 0.43 \, in.$$

O.K, it is less than allowable limit.

If we use the extreme condition and choose $N_{60} = 18$

$$\alpha_2 = \frac{1.71}{N_{60}^{1.4}} = 0.0299$$

$$S_e = 0.14 * (0.0299)(30)^{0.7} \left(\frac{2.896}{2}\right) = 0.065 \, ft = 0.786 \, in. \text{ O.K good}$$

1.11 Check for Overturning:

The ratio of resisting moment to the overturning moment will lead to the factor of safety which as can be noticed is relatively a high number.

$$FS_{Overturning} = \frac{\sum M_R}{\sum M_O} = \frac{3200(15)}{1000} = 48 \quad > 5 \ ok$$

1.12 Sliding Analysis

Assuming no Cohesion on top of the 3rd layer. It is important to note that from passive earth pressure on the side of footing is neglected to be on the safe side.

$$FS_{Sliding} = \frac{\sum V \tan(k_1 \emptyset)}{V} = \frac{(3200) \tan\left(\frac{2}{3}(46)\right)}{100} = 18.97 > 3 \ ok$$

$$FS_{Sliding} = 18.97 > 3 \ good$$

1.13 Bearing capacity

As it was mentioned that settlement is governing factor for design of this footing. At this stage the bearing capacity is calculated in addition to net bearing capacity.

1.14 Modification of bearing capacity ratio for water table

$$\bar{\gamma} = \gamma' + \frac{d}{B}(\gamma - \gamma')$$

$0 < d=19.5 < B=30$

Arzhang Zamani

Layer 3: $\gamma = 120$ pcf

$$\bar{\gamma} = (130 - 62.4) + \frac{19.5}{30}(120 - 130 + 62.4) = 101.66 \text{ pcf}$$

Bearing Capacity using Mayerhof (1963):

$D_f = 5.5 \, ft.$ **Square footing**

Using Layer 2 as a basis, Layer 2:

$\varphi' = 42\,°$

FS = 3

$$q_{all} = \frac{Q_{all}}{B^2} \qquad q_{all} = \frac{q_u}{FS}$$

$N_q = 85.38$ $\qquad\qquad$ $N_\gamma = 155.55$ $\quad$ Table 3.3 $\qquad$ $\gamma = 0.125 \, kcf$

$$q_{all} = \frac{1}{3}\left(qN_qF_{qs}F_{qd} + \frac{1}{2}\gamma BN_\gamma F_{\gamma s}F_{\gamma d}\right)$$

$q = 0.120(5.5) = 0.66 \, kip/ft^2$

$Q_{all} = 3,200 \, kips$

$F_{qs} = 1.9$ $\quad$ $F_{\gamma s} = 0.6$ $\quad$ $F_{qd} = 1 + 2\tan(42)(1 - \sin(42))^2 * \frac{D}{B} = 1.036$

$F_{\gamma d} = 1$

$$q_{all} = \frac{1}{3}\left(qN_qF_{qs}F_{qd} + \frac{1}{2}\gamma BN_\gamma F_{\gamma s}F_{\gamma d}\right) = 95.31\ kip/ft^2$$

$$Q_{all} = q_{all} * 30^2 = 85,779\ kip > 3200\ kip\ O.K$$

$$q_u = \left(qN_qF_{qs}F_{qd} + \frac{1}{2}\gamma BN_\gamma F_{\gamma s}F_{\gamma d}\right) = 285.93\ kip/ft^2$$

$$q_{net} = q_u - q = 285.27\ kip/ft^2$$

As it was expected, the bearing capacity will be easily satisfied with this great size of foundation. Even, the modification of gamma for water table is not very important. Since, the layer 2 has weaker properties than layer 2 and the satisfaction of applied load is appropriate with a good margin.

1.15 Foundation Thickness:

A simplified equation can be derived for individual footing with minimum reinforcement.

Concrete with $f'_c = 3\ ksi$ is used for the foundation. (PCA 1993)

$f'_c = 3\ ksi\ \ W_c = 150\ pcf\ \ \ f_y = 60\ ksi$

Set $\rho = 0.0018(1.11) = 0.002$

Arzhang Zamani

$$R_n = \rho f_y \left(1 - \frac{0.5\rho f_y}{0.85 f'_c} \right) = 117.2 \, psi$$

For 1 ft wide design strip.

$$d^2(req) = \frac{M_u * 1000}{0.9 R_n}$$

$$M_u = q_u \left(\frac{c^2}{2} \right)$$

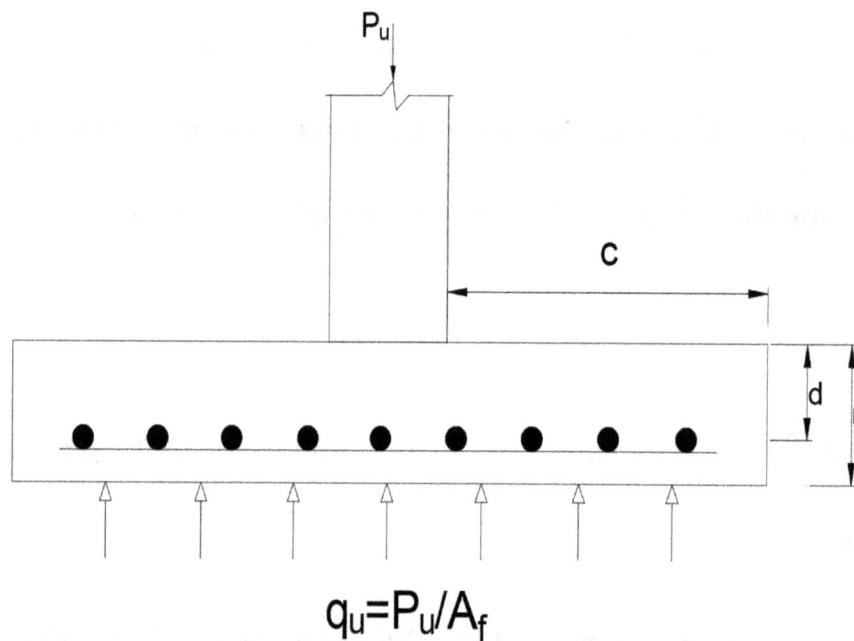

Figure 3 Detailing of a foundation

$$q_u = \frac{P_u}{Af}$$

$$d(req) = 2.2c\sqrt{\frac{P_u}{Af}}$$

$$c = 13.5\,ft.$$

Note:

Simplified approach has been used to determine the reinforcement and thickness of footing.

Normally the strength of foundation is less than other components of structure. For this problem the 3ksi concrete has been used. It is important to mention this point that if we use concrete with higher strength the thickness will be less and consequently the bottom reinforcement will be reduced.

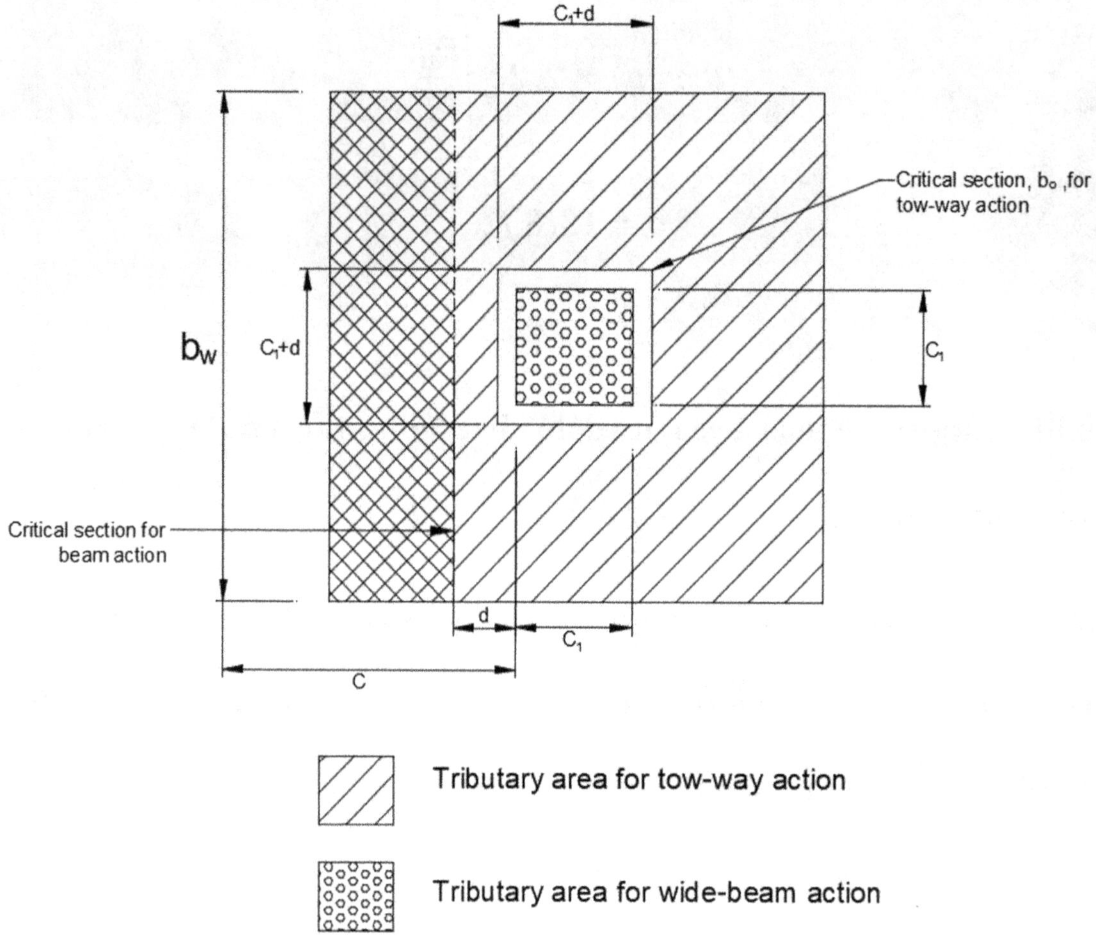
C1+d
Critical section, bo ,for tow-way action
bw
C1+d
C1
Critical section for beam action
d
C1
c
Tributary area for tow-way action
Tributary area for wide-beam action

Figure 4 Critical Sections for shear in footing

One-way shear:

$$V_u = q_u(b_w)(c - d) < 2\emptyset\sqrt{f'_c}b_w d$$

$$\frac{d}{c} = \frac{q_u}{q_u + 2\emptyset\sqrt{f'_c}}$$

Punching shear:

$$V_u < 4\emptyset\sqrt{f'_c}b_o d$$

$$b_o = 4(c1 + d)$$

Therefore,

Thickness: h = d(req) + 4 in > 10 in

$$d(req) = 2.2c\sqrt{\frac{P_u}{Af}}$$

$$c = 13.5\ ft$$

$$d(req) = 2.2(13.5)\sqrt{\frac{3200}{30^2}} = 56\ in$$

h = d (req) + 4 in =60 in > 10 in

d = 56 in

Check one way and punching shear:

One way:

$$V_u = q_u(b_w)(c - d) = \frac{3200}{30^2}(30)\left(13.5 - \frac{56}{12}\right) = 942.22\ kip$$

$$2\emptyset\sqrt{f'_c}b_wd = \frac{2(0.75)\sqrt{3000}30(12)(56)}{1000} = 1656.3 \; kip \; O.K$$

$$V_u < 2\emptyset\sqrt{f'_c}b_wd$$

Punching shear:

$$V_u = q_u(Af - (c1 + d)^2) = \frac{3200}{30^2}\left(900 - \left(3 + \frac{58}{12}\right)^2\right) = 2981.8$$

$$b_o = 4(c1 + d) = b_o = 4(36 + 58) = 376 \; in$$

$$4\emptyset\sqrt{f'_c}b_od = \frac{4(0.75)\sqrt{3000} * 376(56)}{1000} = 3459.9 \; kip$$

$$V_u < 4\emptyset\sqrt{f'_c}b_od \quad O.K$$

Determine the reinforcement:

$$A_s = 0.0018bh = 0.022h \left(\frac{in^2}{ft}\right)$$

$$A_s = 0.0018bh = 0.022(60)\left(\frac{in^2}{ft}\right) = 1.32 \ \frac{in^2}{ft}$$

This is equivalent to # 6 @ 4 in. $f'_c = 3ksi$

Minimum spacing is 18 in. ok

The bars should be placed in both perpendicular directions.

1.16 Geometry and Detailing:

Square foundation with 30 ft. width and 5 ft depth having $f'_c = 3\ ksi$ has been used.

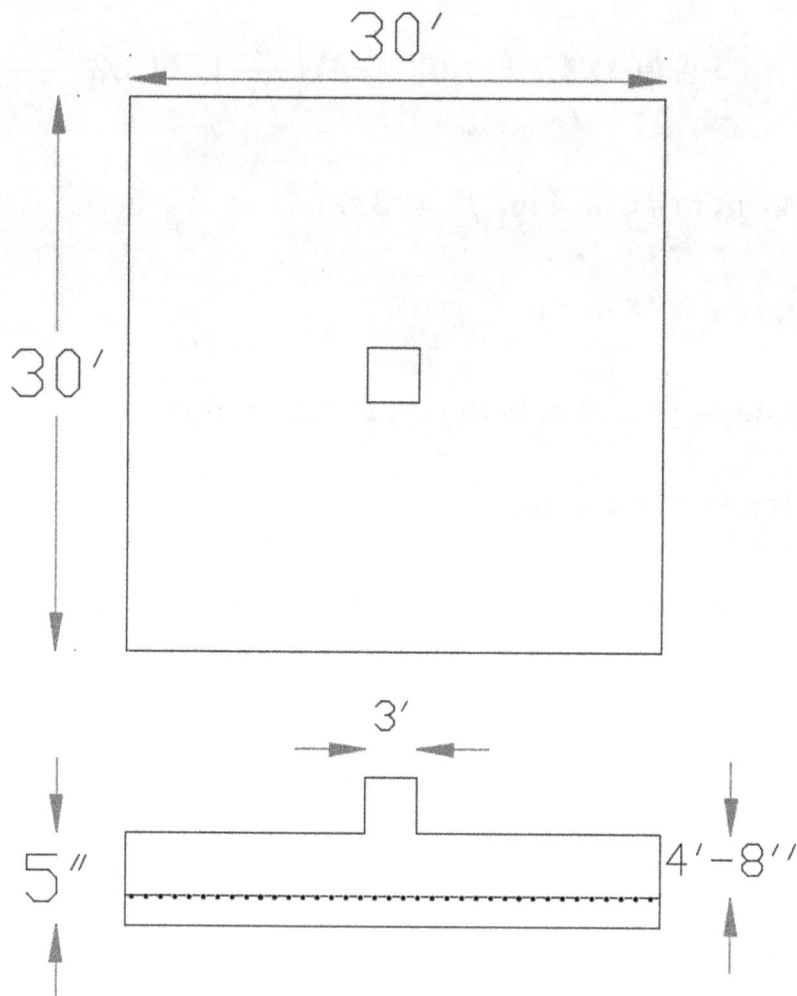

Figure 5 Detailing of a foundation

1.17 Conclusion

In this project the shallow foundation for a bridge pier under given demands have been design in accordance with the latest and most reliable methods of geotechnical engineering.

The analysis is such that the limits that are enforced by the code are satisfied. Settlement according to three methods is checked and bearing capacity is sufficient for the given loads.

Effect of eccentricity is negligible for this problem according to performed analysis in the report. Square foundation with 30 ft. width and 5 ft. thickness having $f'_c = 3\ ksi$ is appropriate for this pier. A simplified approach has been used to determine required thickness of foundation to avoid one way and punching shear.

Chapter 2

Retaining Wall Design

2.1 Problem Statement:

The attached site plan shows part of the overall grading plan for a project. Due to a conflict with right of ways within the site, the civil engineer had to include several retaining walls to resolve the conflict. Your assignment for Project #2 is to design the retaining wall shown within the clouded area shown on the plan. Several soil test borings were drilled as part of the subsurface investigation for the site. Two of those soil test borings (B-1 and B-2) were drilled for that retaining wall. The attached boring logs show the different type of soils encountered at the site including groundwater levels and cave-in depth. The retaining wall should be designed as a geogrid–reinforced retaining wall. Since there will be a roadway and building adjacent to the retaining wall, you should

use a uniform surcharge load of 400 lb/ft$_2$ in your design. The geogrid should be

of the type "TenCate Mirafi" biaxial geogrid and the facing should be

"Keystone" Segmental Block. Information of these products can be found on

the internet at:

http://www.tencate.com/2758/TenCate/Geosynthetics/Region-North-

America/en/en-Geosynthetics---Geotextiles/TenCate-Mirafi-

Products/Geogrids/TenCate-Miragrid-XT

http://www.keystonewalls.com/#

Please check the Appendix for boring data.

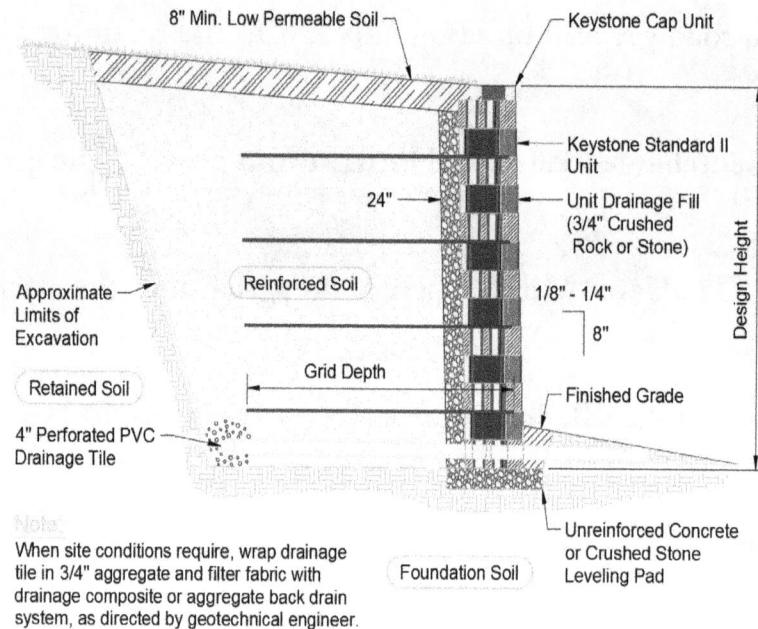

8" Min. Low Permeable Soil — Keystone Cap Unit

Keystone Standard II Unit

Unit Drainage Fill (3/4" Crushed Rock or Stone)

24"

Reinforced Soil

1/8" - 1/4"

8"

Design Height

Approximate Limits of Excavation

Retained Soil

Grid Depth

Finished Grade

4" Perforated PVC Drainage Tile

Note:
When site conditions require, wrap drainage tile in 3/4" aggregate and filter fabric with drainage composite or aggregate back drain system, as directed by geotechnical engineer.

Foundation Soil

Unreinforced Concrete or Crushed Stone Leveling Pad

2.2 "Fundamental Characteristics"

- B1 and B2 boring logs for two sides of wall are determined.

- Geogrid-Reinforcement is the type of retaining wall

- Uniform surcharge for designing is 400 $\frac{lb}{ft^2}$.

- "TenCate Mirafi" biaxial geogrid and the facing should be "Keystone" Segmental Block.

2.3 "Required"

- **Profile of the subsurface stratigraphy including outline of the retaining wall**

- **Size of leveling pad**

- **Type and size of segmental blocks facing used in your design**

- **Type and length of geogrid used**

- **Backfill requirements for the reinforced zone**

- **Vertical spacing of the geogrid**

- **Drainage design beyond typical drainage provisions for segmental block retaining walls**

- **Calculations for internal and external stability excluding calculations for deep seated stability**

- **Cad drawing of section of retaining wall including the drainage provisions**

2.4 Abstract

This project is about designing a retaining wall for given plan in

vicinity of boring logs.

This retaining wall is designed by using geogrid reinforcement. Biaxial Tencate Mirafi is used for geogrid reinforcement. The facing of wall is made of keystone block according to manufacturer's detailing. The demand is uniform surcharge of 400 lb/ft. Drainage system is designed and the fill material is chosen as granular material for sufficient permeability. All of calculations and detailing are provided for this design.

It is very important to add this point, the efficiency of MSE walls can be much higher than conventional retaining walls. It is possible to reach higher heights with more safety and easier installation.

In this project, design for one of applications of this system is explored.

2.5 Phase I: Subsurface Exploration Analysis

It is necessary to describe the mechanical characteristics of soil and types of layers by using provided subsurface exploration tests. Two boring logs on two sides of wall have been given.

The estimate for types of soil layers is:

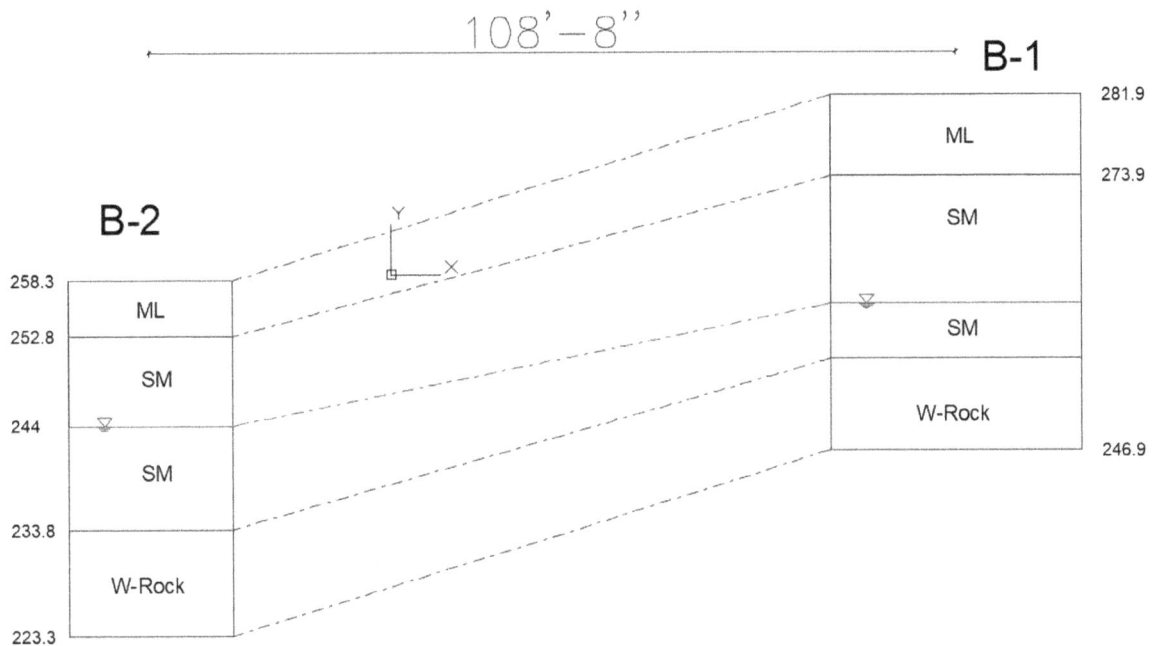

Figure 6 Soil Profile Underneath Boring logs

Note: The scale of plan is 1:360. In other words, one inch is equivalent to 30 ft. in horizontal distance. The height is selected according to given values in contours. As can be noticed from this view that ground water is not an issue for this retaining wall. Therefore, by choosing sufficient fill materials in the back of wall and providing adequate drainage system. We can make sure that water can be drained from the wall in an emergency. The characteristics of soil layer beneath the retaining wall is delivered in the bearing capacity of retaining wall.

2.6 Profile of the subsurface stratigraphy including outline of the retaining wall

A parallel line near B1-B2 which is perpendicular to the wall has been chosen. According to given topography, profile of soil is derived as below.

Proposed Retaining Wall Position:

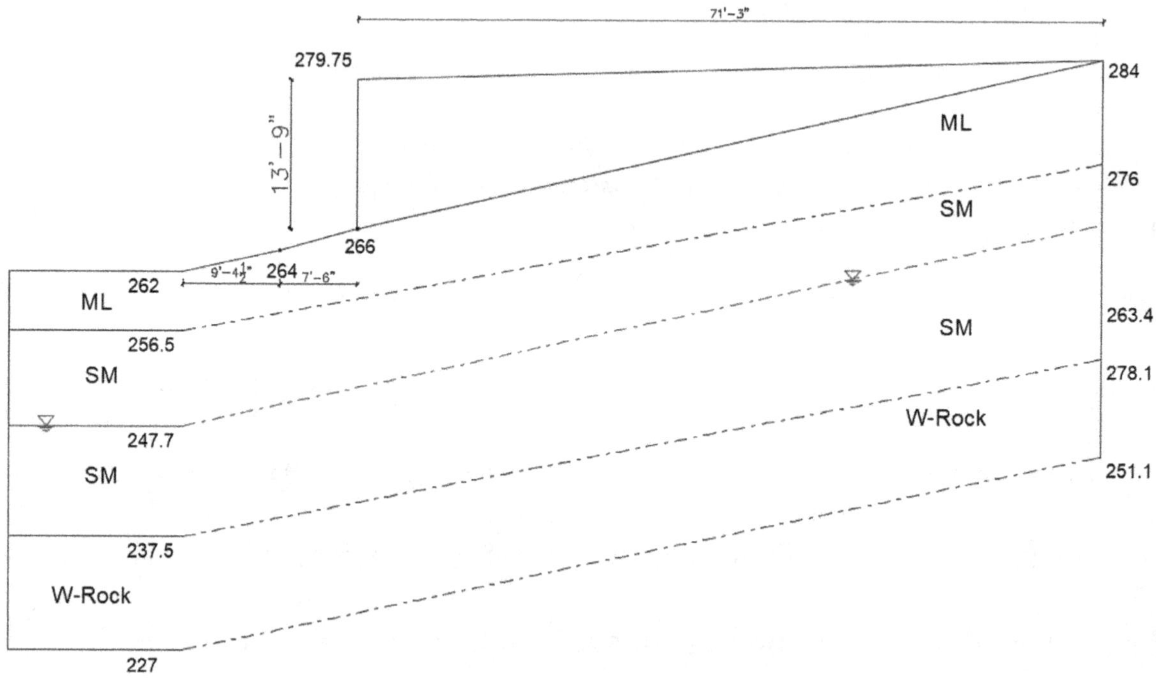

Figure 7 Expected Soil profile in the lie perpendicular to the wall next to B1 and B2

The asked height of wall is 13'-9''. The final height will be chosen consequently according the height of keystone blocks. Therefore, the final height of wall is 14 ft.

It can be noticed that ground water table is not problematic for this retaining wall. By using granular fill and sufficient drainage system and correct installation of pipe and retaining wall by contractor. This wall can drain the probable stream from its wall.

It is important to mention that the difference between the elevation of wall to upper hand ground at about 71 ft. distance is 4 ft. Approximately the slope of that line which connects the 280 elevations to 284 is 5%.

Note: In the calculations, it is assumed to have the vertical force due to surcharge downward and effect of slope is neglected. Thus, this slop is very low, and it does not affect the calculations.

2.7 Geogrid-Reinforcement selection:

TenCate Mirafi® BXG biaxial geogrids ("Tencate Mirafi") are used for base course reinforcement and soil stabilization applications. They offer

Arzhang Zamani

high strength at low strain and are designed for maximum bearing capacity
and shear resistance. TenCate Mirafi® BXG geogrids are constructed of
high tenacity, high molecular weight woven polyester to deliver increased
passive bearing resistance. Coated with a polymer coating, TenCate
Mirafi® BXG geogrids provide optimum interaction in all soil types. It is
for soil reinforcement. (TENCATE, TENCATE GEOsynthetics, 2015)

TenCate Mirafi BXG11 because it has the largest Ultimate Tensile Strength
Ultimate Tensile Strength = 2500 lbs. /ft. (TENCATE, TENCATE MIRAFI
2015)

Mechanical Properties of TenCate Mirafi BXG11:

Table 3 Properties of Geogrid

Mechanical Properties	Test Method	Unit	Minimum Average Roll Value	
			MD	CD
Tensile Strength (at ultimate)	ASTM D6637	lbs/ft (kN/m)	2500 (36.5)	2500 (36.5)
Tensile Strength (at 1% strain)	ASTM D6637	lbs/ft (kN/m)	375 (5.5)	375 (5.5)
Tensile Strength (at 2% strain)	ASTM D6637	lbs/ft (kN/m)	625 (9.1)	625 (9.1)
Tensile Strength (at 5% strain)	ASTM D6637	lbs/ft (kN/m)	1000 (14.6)	1000 (14.6)
Tensile Modulus (at 1% strain)	ASTM D6637	lbs/ft (kN/m)	37500 (547)	37500 (547)
UV Resistance (at 500 hours)	ASTM D4355	% strength retained	70	

Figure 8 MIRAFI BXG11 Geogrid Reinforcement (Biaxial)

The good point about biaxial geogrid is that, the contractor may not make mistake by placing it in a wrong direction. In case of uniaxial geogrids direction of installation must be checked precisely for correct behavior of system.

2.8   Wall type Selection:

Use Keystone Straight Split Standard I.

The drawing and details from manufacturers are provided below and in the final drawings.

Series I

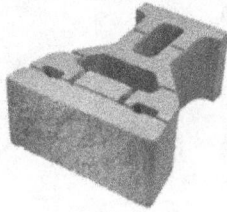

Height	Width	Depth	Weight
8"	18"	18-21"	102-114 lbs
203mm	457mm	457-533mm	46-52 kg

UNIT COLOR, DIMENSIONS, WEIGHT & AVAILABILITY VARIES BY MANUFACTURER.

Units/Sq.Ft.

1

Figure 9 Keystone Straight Split Standard I

Note: The height of blocks will govern the finished height and vertical spacing of geogrid.

Therefore, the density of the block is: $\left(\frac{108}{\left(\frac{18}{12}\right)\left(\frac{18}{12}\right)\left(\frac{8}{12}\right)}\right) = 72 \, pcf$

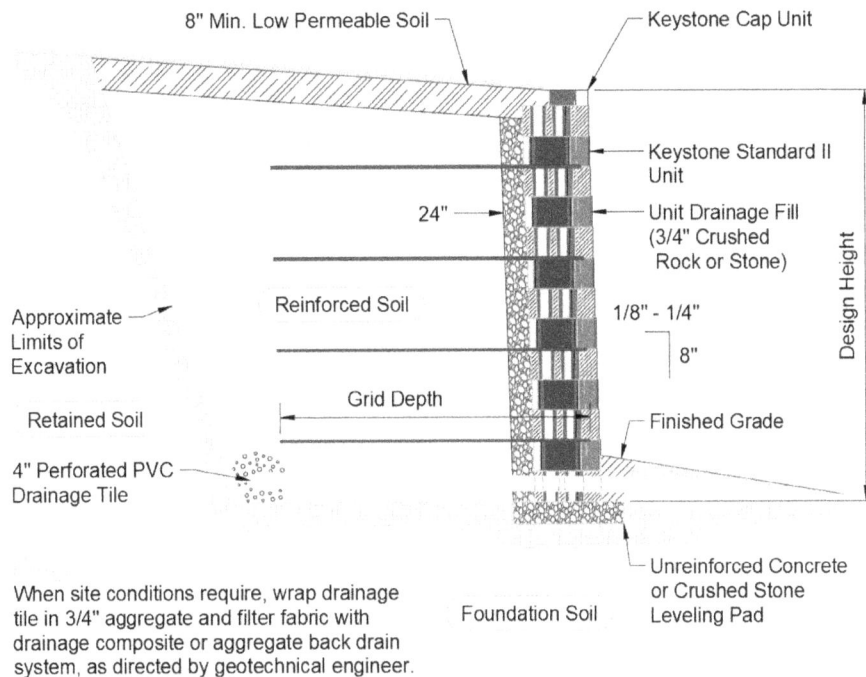

Figure 10 Typical retaining wall

This detailing is going to be used for this problem. As can be seen this configuration accompanied by correct installation and following the notes will result is appropriate retaining wall for this application. The important part of retaining wall correct installation of drainage pipe and correct choosing of backfill properties and detailing. Therefore, during the perspiration the water will drain easily through the pipe. (KESTONE 2015)

Important note about the spacing of geogrid is that, it must be between the heights of each block.

Arzhang Zamani

2.9   Size of leveling pad

According to Keystone manual the size of leveling pad is selected to be compatible with the facing and retaining wall.

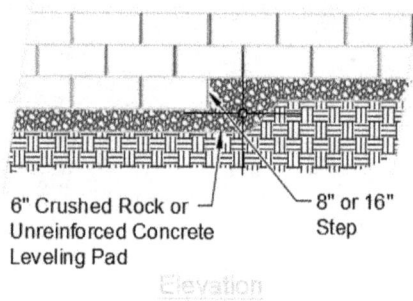

6" Crushed Rock or
Unreinforced Concrete
Leveling Pad

8" or 16"
Step

Elevation

Note:

1. The leveling pad is to be constructed of crushed stone or 2000 psi ± unreinforced concrete.

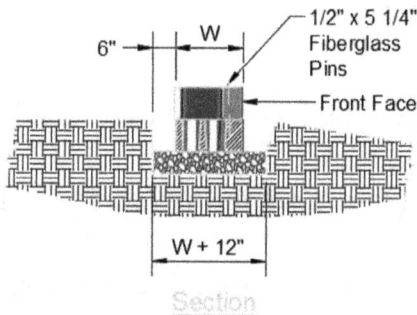

6"   W   1/2" x 5 1/4"
Fiberglass
Pins

Front Face

W + 12"

Section

Leveling Pad Properties:

Min thickness = 6" on either side of wall.
Width of pad = 12" + 18" = 30".
Min Height = 6"

Figure 11 Leveling Pad Detail

56

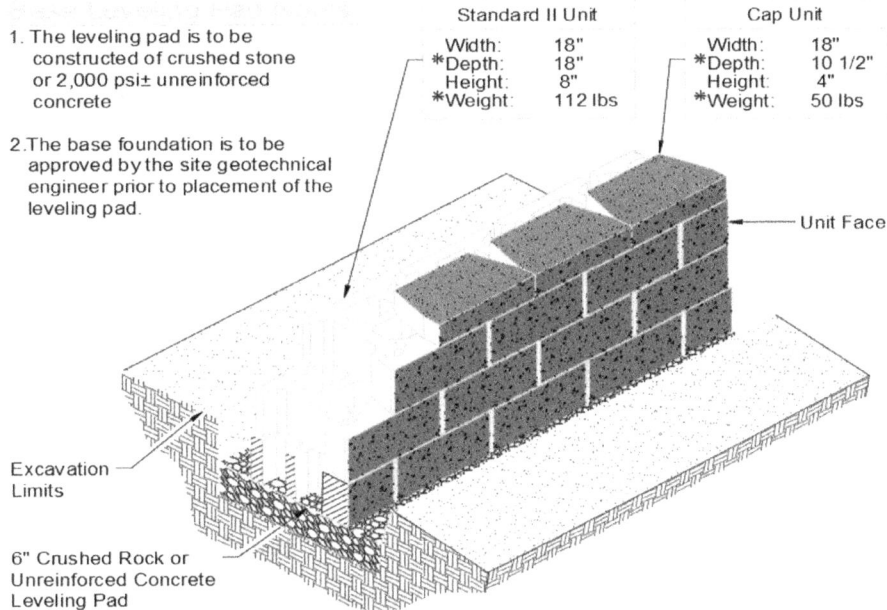

1. The leveling pad is to be constructed of crushed stone or 2,000 psi± unreinforced concrete

2. The base foundation is to be approved by the site geotechnical engineer prior to placement of the leveling pad.

Standard II Unit
Width: 18"
*Depth: 18"
Height: 8"
*Weight: 112 lbs

Cap Unit
Width: 18"
*Depth: 10 1/2"
Height: 4"
*Weight: 50 lbs

Unit Face

Excavation Limits

6" Crushed Rock or Unreinforced Concrete Leveling Pad

Figure 12 Base leveling pad notes and details

Problem Statement

Height of wall = 13.75 ft.
Demand (Surcharge) = 400 lb/ft. uniform surcharge.

Calculate Total Number of Blocks

$$Number\ of\ blocks = \frac{Total\ Wall\ H}{Block\ H} = \frac{13.75*12}{8} = 20.6\ \ use\ \ 21\ blocks$$

Height of wall = 21*(8/12) = 14 ft.

According to elevation view and height of wall. It is a fill situation. It means that we must choose granular soil in the back of wall for fil materials.

Arzhang Zamani

Select Soil for Fill

Use Granular soil, uniform sand.

$$(C' = 0 \quad \varphi' = 30° \gamma = 110 \text{ lb/ft}^3)$$

2.10 Mechanical Properties:

Rankine active pressure:

Step 1:

$$K_a = tan^2\left(45 - \frac{\emptyset}{2}\right) = tan^2(30) = \frac{1}{3}$$

$$Z = 14 \text{ ft.} \quad \& \gamma = 110 \text{ lb/ft}^3)$$

$$\gamma Z K_a = 110(14)\left(\frac{1}{3}\right) = 513.33 \quad \frac{\text{lb}}{\text{ft}^2}$$

Therefore,

$$\sigma'_a = 513.33 \frac{\text{lb}}{\text{ft}^2}$$

$$T_{all} = \frac{T_{ult}}{RF_{id} \times RF_{cr} \times RF_{cbd}}$$ **(Das 2010)**

*Average value of recommended reduction factor has been chosen.

$RF_{id} = \frac{1.1+1.4}{2} = 1.25$ **(Reduction factor for installation damage)**

$RF_{cr} = \frac{2+3}{2} = 2.5$ **(Reduction factor for creep**

$RF_{cbd} = \frac{1.1+1.5}{2} = 1.3$ **(Reduction factor for chemical and biological degradation)**

58

$$T_{all} = \frac{2500}{(1.25)(2.5)(1.3)} = 615.38 \frac{lb}{ft}$$

$$S_V = \frac{T_{all}C_r}{\sigma'_a FS_{(B)}}$$

Assume C_r=1.0 because no spacing between pieces of geogrid is considered.

$$FS_{(B)} = 1.5$$

$S_V = \frac{T_{all}C_r}{\sigma'_a FS_{(B)}} = \frac{615.38(1)}{513.33(1.5)} = 0.79 \, ft \; \text{USE 1}$

Therefore, $S_V = 1 \, ft$.

NOTE:

According the limitation of inserting the geogrids between heights of blocks.

Since, the height of each block is 8 in. 3(8) =24in=2ft is a practical way to insert

geogrids.

USE 2' spacing.

$$S_V = 2'$$

Arzhang Zamani

H=14ft

$$L = l_r + l_e$$

$$l_r = \frac{H - z}{\tan\left(45 + \frac{\phi_1'}{2}\right)}$$

$$l_r = \frac{14}{\tan(60)} = 8.08 \ ft$$

$\varphi' = 30°$ C_r=1.0

C_i=0.75 (Well graded Sand has been used for back fill)

$$l_e = \frac{S_V K_a \, FS_{(P)}}{2C_r C_i \tan \phi_1'}$$

$$FS_{(p)} = \frac{Resistance\ to\ pullout\ at\ the\ given\ normal\ effective\ stress}{Pullout\ force}$$

$$FS_{(p)} = 1.5$$

$$l_e = \frac{(2)\left(\frac{1}{3}\right) 1.5}{2(0.75)\tan(30)} = 1.1547 \ ft$$

$$L = l_r + l_e = 8.08 + 1.1547 = 9.23 \ ft$$

USE L=10 ft.

Checking the external stability for all other steps.

External Stability Checks

60

Step 9)

SLIDING:

$$FS_{(sliding)} = \frac{(W_1 + W_2 + \cdots + qa')(\tan(k\varphi'))}{P_a}$$

$$P_a = \frac{1}{2} * (\gamma)(H)^2 K_a = (0.5)(110)(14)^2 \left(\frac{1}{3}\right) = 3594 \frac{lb}{ft}$$

$$FS_{sliding}$$
$$= \frac{[(weight\ of\ soil) + (weight\ of\ wall) + (overburden\ pressure)]\left(\tan(\frac{2}{3}\varphi)\right)}{P_a}$$

$$FS_{sliding} = \frac{[(10*14*110) + (72)(14)(1.5) + (400*10)](0.364)}{3594} = 2.11$$

Not good. The factor of safety must be greater than 3. The length of geogrid must be modified.

Therefore,

$$FS_{sliding} = \frac{[(L*14*110) + (72)(14)(1.5) + (400*L)](0.364)}{3594} > 3$$

Solve for L: L = 14.49 ft.   USE L=15 ft.

Arzhang Zamani

$$FS_{sliding} = \frac{[(L * 15 * 110) + (72)(14)(1.5) + (400 * 15)](0.364)}{3594}$$

$$= 3.1 \ > 3 \ ok$$

Step 10)

Overturning:

$$FS_{overturning} = \frac{M_R}{M_o}$$

$$FS_{overturning} = \frac{\left(15 * 110 * \left(\frac{15}{2}\right)\right) + \left(400 * 15 * \left(\frac{15}{2}\right)\right)}{\left(3594 * \left(\frac{14}{3}\right)\right)} = 3.42 > 3$$

OK

Step 11)

Check for ultimate bearing capacity.

To check for ultimate bearing capacity. Boring log 2 (B-2) is used for finding characteristics of that layer.

$$\eta_h = 79\% \ Hammer \ Efficiency$$

$$\eta_B = 1 \ \eta_s = 1 \ \eta_R = 1$$

According to Table 6 (NAVFAC DESIGN MANUAL 7.01)

Unit weight of ML (Silty clay with stone or rock fragments) is

$$\frac{115+151}{2} = 133\,\frac{lb}{ft^3}$$

It is a good approximation for value of γ.

Layer 1: (0-5.5ft)

(N)ave = 10

$$N_{60} = \frac{N_{ave}(\eta_h)(\eta_B)(\eta_s)(\eta_R)}{60} = \frac{10(79)(1)(1)(1)}{60} = 13$$

(Kulhawy and Mayne, 1990)

$$\sigma'_0 = 133\left(\frac{5.5}{2}\right) = 365.75\ pcf \quad P_a = 2,000\ pcf$$

$$\varphi' = \tan^{-1}\left(\frac{N_{60}}{12.2 + 20.3\left(\frac{\sigma'_0}{P_a}\right)}\right)^{0.34}$$

$$\varphi' = \tan^{-1}\left(\frac{13}{12.2 + 20.3\left(\frac{365.75}{2000}\right)}\right)^{0.34} = 43°$$

Arzhang Zamani

Therefore,

$$\gamma(2) = 133\,\frac{lb}{ft^3}$$

$$\varphi'(2) = 43\,°$$

Then,

$$N_\gamma = 186.54$$

$$q_{ult} = (0.5)(\gamma)(L)\big(N_\gamma\big) = 0.5(133)(15)(186.54) = 186.074\,\frac{kip}{ft^2}$$

$$q + \gamma H = 400 + 110(14) = 1940\,lb/ft^2$$

$$FS_{bearing} = \frac{(0.5)(\gamma)(L)\big(N_\gamma\big)}{q + \gamma H} = \frac{0.5(133)(15)(186.54)}{400 + (110)(14)} = 95 > 3$$

OK

Since (Kulhawy and Mayne, 1990) gave us a high friction angle. The conservative assumption is that choose smaller value.

Assume $\varphi'(2) = 30\,°$

Then,

$$N_\gamma = 22.4$$

$$q_{ult} = (0.5)(\gamma)(L)(N_\gamma) = 0.5(133)(15)(22.4) = 22,344 \ \frac{kip}{ft^2}$$

$$q + \gamma H = 400 + 110(14) = 1940 \ lb/ft^2$$

$$FS_{bearing} = \frac{(0.5)(\gamma)(L)(N_\gamma)}{q+\gamma H} = \frac{0.5(133)(15)(22.4)}{400+(110)(14)} = 11.52 > 3 \ \ OK$$

By adjusting and decreasing the friction angle the factor of safety becomes more reasonable. It is also greater than 3 and acceptable.

Therefore, the Internal and external stability of retaining wall with geogrid are satisfied.

2.11 Drainage Check

From Keystone, 1 cubic foot of drainage fill required for every square foot of wall.

Because this is a fill situation, and the ground water table is below the wall and leveling pad, a 4" radius slotted PVC pipe is required along with drainage tiles. This specification is from the Keystone Specifications page. Low permeable soil must be placed on top of retaining wall surface. The 4" radius slotted PVC pipe is placed to collect probable accumulated water during perspiration. In this case water will be drained from the back of wall

Arzhang Zamani

very soon. The effect of stuffiest drainage system is crucial for making sure that retaining wall performance is appropriate.

2.12 CAD Detailing

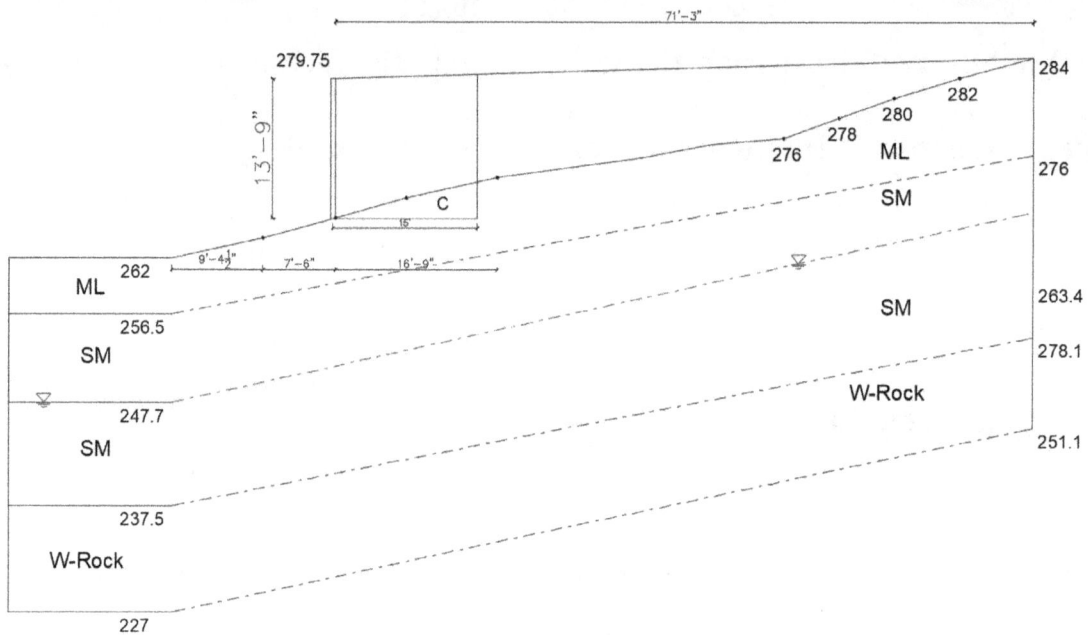

Figure 13 Soil profile, wall position, and cut volume

As can be noticed, we have a fill situation for most of the parts, the cut volume per linear foot is: $27 ft^2(1) = 27$ ft^3. The Figure 13 shows the area for cut.

The retaining wall with keystone facing and using geogrid reinforcement is designed. The length of geogrid is 15 ft. and the vertical spacing of geogrids are one ft. Geogrid cover all of surface. The backfill is chosen from well graded granular soil to have sufficient drainage capability. The drainage pipe is 4 in. in diameter and the location of it is shown in CAD details. "Tencate Mirafi" geogrid is used, which has high tensile capacity in both directions (biaxial). Therefore, we are not worry about the wrong installation of that on the surface.

Appendix

Appendix includes boring data, final detailing, and catalog for used material in this project.

PROJECT NAME:	PROJECT NUMBER: 123	BORING NUMBER: B-1
CLIENT NAME:	CLIENT PROJECT NUMBER:	SHEET 1 OF 2
DRILLER: JC	LOCATION: As Staked	ELEVATION (FEET): 281.9'

WATER LEVELS	DATE	TIME	DEPTH	CAVED	
ENCOUNTERED:	3/15/12	2:50 PM	20.6'		DATE START: 3/15/12
BEFORE CASING PULLED:	3/15/12	4:10 PM	26.2'		DATE FINISH: 3/16/12
AFTER CASING PULLED:	3/15/12	4:35 PM	22.4'	30.4'	METHOD: 3 1/4" ID HSA
					EQUIPMENT USED: CME 45C
LONG TERM:	3/16/12	3:45 PM	18.4'	22.5'	REVIEWER: CT

ELEVATION (FEET)	LEGEND	USCS	CLASSIFICATION	DEPTH (FEET)	SAMPLE #	BLOWS PER 6 INCHES	N VALUE	RECOVERY (INCHES)	MOISTURE CONTENT	REMARKS
281.9 281.7		T	TOPSOIL: 3"	0						
		ML	Stiff yellowish brown and reddish brown sandy SILT with rock fragments, moist		1	2-4-5	9	16		
				5	2	4-3-4	7	12		
					3	5-5-6	11	18		
273.9				10	4	6-7-8	15	18		
		SM	Medium dense mottled brown, red, black and pink silty fine to coarse SAND with rock fragments, moist	15	5	8-9-9	18	18		
				20	6	9-7-9	16	18		
				25	7	8-10-9	19	18		
255.9										
		W	WEATHERED ROCK, sampled as mottled brown, red, black, and pink silty gravel with	30	8	23-36-30	66	18		

Notes:

PROJECT NAME:	PROJECT NUMBER:	BORING NUMBER:		
	123	B-1		
CLIENT NAME:	CLIENT PROJECT NUMBER:	SHEET 2 OF 2		

ELEVATION (FEET)	LEGEND	USCS	CLASSIFICATION	DEPTH (FEET)	SAMPLE #	BLOWS PER 6 INCHES	N VALUE	RECOVERY (INCHES)	MOISTURE CONTENT	REMARKS
			sand, moist							
246.9				35	9	28-30-42	72	18		
				40						Boring terminated at 35 feet
				45						
				50						
				55						
				60						
				65						
				70						

PROJECT NAME:			PROJECT NUMBER: 123			BORING NUMBER: B-2			
CLIENT NAME:			CLIENT PROJECT NUMBER:			SHEET 1 OF 2			
DRILLER: JC			LOCATION: As Staked			ELEVATION (FEET): 258.3'			
WATER LEVELS			DATE	TIME	DEPTH	CAVED	DATE START: 3/16/12		
ENCOUNTERED:			3/16/12	1:26 PM	14.3'		DATE FINISH: 3/17/12		
BEFORE CASING PULLED:			3/16/12	3:18 PM	18.3'		METHOD: 3 1/4" ID HSA		
AFTER CASING PULLED:			3/16/12	4:27 PM	20.1'	26.0'	EQUIPMENT USED: CME 45C		
LONG TERM:			3/17/12	12:45 PM	11.8'	14.2'	REVIEWER: CT		

ELEVATION (FEET)	LEGEND	USCS	CLASSIFICATION	DEPTH (FEET)	SAMPLE #	BLOWS PER 6 INCHES	N VALUE	RECOVERY (INCHES)	MOISTURE CONTENT	REMARKS
258.3 257.8		T	TOPSOIL: 6"	0						
		ML	Firm to stiff mottled yellow, brown, white, gray and red sandy SILT with rock fragments, moist		1	3-3-5	8	18		
					2	6-6-5	11	16		
252.8				5						
					3	3-8-9	17	18		
					4	7-7-9	16	12		
				10						
		SM	Medium dense mottled brown, red, black and pink silty fine to coarse SAND with rock fragments, moist	15	5	10-10-9	19	14		
					6	11-11-10	21	18		
				20						
233.8					7	8-12-50/4	72	18		
				25						
		W	WEATHERED ROCK, sampled as mottled brown, red, black, and pink silty gravel with sand, moist	30	8	36-38-40	78	18		

Notes:

TEST BORING LOG

PROJECT NAME:	PROJECT NUMBER:		BORING NUMBER:		
	123		B-2		
CLIENT NAME:	CLIENT PROJECT NUMBER:		SHEET	2	OF	2

ELEVATION (FEET)	LEGEND	USCS	CLASSIFICATION	DEPTH (FEET)	SAMPLE #	BLOWS PER 6 INCHES	N VALUE	RECOVERY (INCHES)	MOISTURE CONTENT	REMARKS
					9	25-31-29	60	18		
223.3				35						Boring terminated at 35 feet
				40						
				45						
				50						
				55						
				60						
				65						
				70						

KEY TO SYMBOLS

Symbol   Description

<u>Strata symbols</u>

[symbol]   Topsoil

[symbol]   Silt

[symbol]   Silty Sand

[symbol]   Weathered Rock

<u>Misc. Symbols</u>

[symbol]   Water level during
                drilling

[symbol]   Long term water reading

<u>Notes:</u>

1. Exploratory soil test borings were drilled between March 15 and
   March 17, 2012 using 3 1/4" inside-diameter hollow stem augers.

2. Groundwater measurements are recorded on the logs when encountered,
   and after a stabilization period.

3. Boring locations were taped from existing features and elevations
   extrapolated from the final design schematic.

4. These logs are subject to the limitations of visual classifications and
   the interpretations of the geotechnical engineer.

5. Results of the tests conducted on samples recovered are reported on the
   logs.

q=400lb/ft

14'

264

ML

SM

Keystone Cap Unit

Keystone Standard Unit

Unit Drainage Fill (3/4" Crushed Rock or Stone)

Finished Grade

8"

1/8" - 1/4"

Unreinforced Concrete or Crushed Stone Leveling Pad

Foundation Soil

266

Note
When site conditions require, wrap drainage tile in 3/4" aggregate and filter fabric with drainage composite or aggregate back drain system, as directed by geotechnical engineer.

280

8" Min. Low Permeable Soil

24"

Sv=2'

2'

Retained Soil

15'

Approximate Limits of Excavation

4" Perforated PVC Drainage Tile

References

The following references were used to create this book and can be utilized as a reference for this book.

- AASHTO LRFD Bridge Design Specifications, 7th Edition with 2015 Interim Revisions. Published by the American Association of State Highway and Transportation Officials (AASHTO), Washington, D.C.

- ACI 318-14 (2014). "Building Code Requirements for Reinforced Concrete" and "Commentary on Building Code Requirements for Reinforced Concrete'" American Concrete Institute, P.O. 4754 Redford Station, Detroit, Michigan 48912. Phone: (248) 848-3700, Fax: (248) 848-3701.

- Foundations & earth structures: Design manual 7.2. (1986). Alexandria, VA: U.S. Navy, Naval Facilities Engineering Command.

- Portland Cement Association, (1993). Simplified Design: Reinforced Concrete Buildings of Moderate Size and Height (2nd Edition).

- "Bridge Engineering Handbook," Wai-Faf Chen and Lian Duan, 1999, CRC Press

- Das, Braja M (2010). Principles of geotechnical engineering (7th ed). Cengage

  Learning, Stamford, CT

- KESTONE, (2015). http://www.keystonewalls.com. April 20.

  http://www.keystonewalls.com

- TENCATE, (2015). TENCATE Geosynthetics. April 20.

  http://www.tencate.com/amer/geosynthetics/products/geogrids/TenCate-Mirafi-

  BXG/default.aspx.

- TENCATE MIRAFI, Specifications, (2015)

  https://www.tencategeo.us/media/712345ea-673c-4423-83f9-

  982d77faa682/LkcFQQ/TenCate%20Geosynthetics/Documents%20AMER/Techn

  ical%20Data%20Sheets/Woven/Mirafi%20RSi-Series/TDS_RS380i%20161026.pdf

www.ingramcontent.com/pod-product-compliance
Lightning Source LLC
Chambersburg PA
CBHW051231200326
41519CB00025B/7331